Beginning Calculator Activity Book

Problem Solving Using Four-Function Calculators

by
Carol A. Thornton,
Cynthia W. Langrall,
Roger Day,
and
Graham A. Jones

GRADES 1 - 4

Table of Contents

Section 3: **Measurement**

Section 4: **Data and Chance**

Introduction

The *Beginning Calculator Activity Book* contains over 30 problem-solving tasks for children in grades 1 through 4. These tasks challenge children to explore and solve problems in four, key mathematical areas:
- Number Sense
- Patterns
- Data and Chance
- Measurement

This resource book supplements your mathematics program and can be used in regular, special education, and remedial classrooms. The mathematical topics and the spirit of the problem-solving tasks are consistent with recommendations in the *NCTM Curriculum and Evaluation Standards* and the *NCTM Addenda Series,* Grades 1 - 4. The problem-solving tasks incorporate:
- estimation
- mathematical reasoning
- natural connections for using the Basic Calc-U-Vue or other four-function calculators
- written and oral communication

Unless otherwise suggested, it is recommended that children work collaboratively with a partner as they engage in the activities of this book. An opening question sets the stage for each activity — inviting students to predict or estimate the outcome of a problem task. The activities engage students in completing tasks which utilize a calculator as a natural tool, and invite children to communicate their results in writing or by orally sharing their ideas. Children are often asked to reflect on their original predictions or estimations, and to compare these with actual outcomes. Sharing and discussion helps children solidify understandings and find new applications for the ideas presented.

The problem-solving activities in this book follow a flexible format with the following features:

Work Together: Children explore and complete a problem task while working together with a partner.

Write About It: Children record their thinking or explain their solution approach in a math journal or notebook. Students can also write about their math experiences on the back of the activity sheets. Students can keep their math journal pages in a binder.

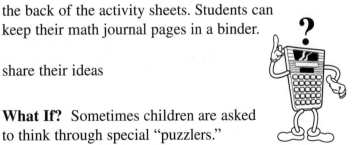

Talk About It: Children share their ideas with others in the class.

What If? Sometimes children are asked to think through special "puzzlers."

A Step Beyond: From time to time, children are challenged to extend their thinking about a problem task.

On some pages, students are reminded of Activity Masters and special materials needed to carry out an activity. Before beginning an activity, check the bottom of the page for the following icon:

It is assumed that all activities require a calculator, paper and pencil.

Teaching Notes are provided at the beginning of each of the four sections. These include an overview of the section as well as teaching suggestions for each activity.

USING CALCULATORS:

There are a wide variety of four-function calculators readily available today. In addition to the common [+] [−] [×] [÷] [=] keys, the following features are common to most:

 On/Clear: Turns the calculator on and then clears the display, constant (if any), and arithmetic operation currently stored in the calculator.

 Clear/Clear Entry: Some calculators combine the C and CE functions on one key. Press once to perform "Clear Entry." Press twice consecutively to perform "Clear."

 Clear Entry: Clears the display only.

 All Clear: Some calculators also have this key, which clears the display, the memory, any constant, and any arithmetic operation.

 Add to Memory: Adds the number currently on the display to the memory. The memory is cumulative.

 Subtract from Memory: Subtracts the number on the display from the number in memory.

 Memory Recall/Memory Clear: Press once to recall to the display window the number in memory. Press twice to clear information in memory.

Section 1: **Patterns**

OVERVIEW

A variety of number patterns and relationships are explored in this section. Students engage in tasks that require them to recognize, describe, and extend patterns, as well as make predictions based on their interpretations and analysis of patterns. In many activities, the calculator functions as a tool for generating counting patterns.

TEACHING NOTES

What's the Pattern? Students use the calculator to explore skip counting patterns. This activity is one of many that utilize the calculator's constant function. While most 4-function calculators will have a constant function (sometimes called a "hot equals"), the keystrokes given for this activity might not apply to all calculators. Some calculators require the following keystrokes:

Refer to the section on constant functions in your calculator's manual for specific details.

What's Next? The calculator's constant function is used to examine the concepts of "one more" and "one less" when counting on or counting back from a given number.

Eyes Closed, Eyes Open: In this partner game, students use the calculator to check as they mentally count on or count back from a given number.

100's Chart: Students use the calculator to skip count by 2, recording different counting patterns on the 100's Chart. They also write a brief description of each pattern. Instead of placing counters on the 100's Chart as suggested in the activity, patterns can be colored directly onto the chart. Concepts of odd and even should emerge as students compare and contrast the different patterns they generate.

Chart Patterns: In this activity, students examine a sequence or block of numbers on the 100's Chart to find patterns in the sums of selected numbers. For example, in any "line" of three numbers, the "middle" number is one half the sum of the other two numbers. Have students calculate the "Diagonal Sums" to see that they are equal. Have students add the digits that make up numbers in columns. Ask them to identify the patterns they see. Students will identify various other patterns as they examine 3x3 and 4x4 blocks of numbers. Ask them why these patterns work. A 200's Chart has been included to encourage the exploration of three-digit numbers.

Guess My Rule ⊞ ⊟: In this activity, one student uses the constant function "in secret" to enter a counting on or counting back rule into the calculator. Another student must identify the counting pattern to "guess the rule." For students who require greater structure, provide recording sheets like the following:

Game 1		Game 2		Game 3	
Enter	Display	Enter	Display	Enter	Display
———	———	———	———	———	———
———	———	———	———	———	———
———	———	———	———	———	———
———	———	———	———	———	———
Rule: ————		Rule: ————		Rule: ————	

Odd or Even? Concepts of odd and even are explored as students use the calculator to test their conjectures about the sums and differences of odd and even numbers.

My Turn, Your Turn ⊞ ⊟: Students continue to explore odd and even numbers. This activity features a fun way of randomly selecting a number on the calculator—with eyes closed, randomly punching one or more digit keys.

Evens vs. Odds: This time, students examine the effects of multiplication on odd and even numbers. The *Talk About It* section of the activity will allow you to assess students' intuitive understandings of these concepts.

My Turn, Your Turn ⊠: Students continue to explore odd and even numbers using multiplication.

Guess My Rule ⊠ ⊡: The operations of multiplication and division are used to generate "rules" similar to the *Guess My Rule*⊠ ⊡ activity. Be aware that some calculators may require different keystrokes to set the constant function for multiplication and division. Consult the manual for your calculator or try the following:

For Example: To multiply by 5, enter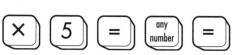

In **A Step Beyond**, students generate "divided by" rules. Decimal displays are certain to occur and will have to be interpreted by students as they identify a pattern to "Guess the rule."

Patterns with Zeros: This activity can be conducted with the whole class. First, direct students to use their calculator to complete the top part of the activity sheet. Keep a tally of the number of seconds each child takes to complete this section. Ask students to describe any patterns they see. Now instruct the class to complete the bottom part of the sheet without the calculator. Suggest that they look for a pattern to use. Keep a tally of the number of seconds each child takes to complete this section. Discuss whether it was faster to use the calculator or to compute mentally. Encourage students to explain why the calculator isn't always the fastest or most efficient way to compute.

What's the Pattern?

Can you tell what number comes next?

Work Together

- Use your 🧮 to count.
- Watch the display window each time you press ⌗=⌗.

☁ Press = a few more times.

ON/C 1 + 2 = = = ...

ON/C 3 0 + 5 = = = ...

ON/C 2 0 + 1 0 = = = ...

Write About It

What different patterns do you see?

What does the = button do?

Talk About It

Talk about your patterns.

What number would come next in each count?

What's Next?

Can you tell what number comes next?

Work Together

- Use your [calculator] to count.

- Watch the display window each time you press $=$.

A.
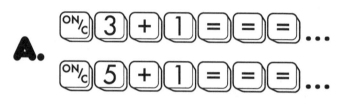
$$\text{ON/C} \; 3 \; + \; 1 \; = \; = \; = \; \dots$$
$$\text{ON/C} \; 5 \; + \; 1 \; = \; = \; = \; \dots$$

B.
$$\text{ON/C} \; 1 \; 0 \; - \; 1 \; = \; = \; = \; \dots$$
$$\text{ON/C} \; 8 \; - \; 1 \; = \; = \; = \; \dots$$

Write About It

What's the pattern for A?

What's the pattern for B?

Talk About It

Talk about your patterns.

What number would come next in each count?

Eyes Closed, Eyes Open

Can you count to the target?

Work Together

- Use your  to count.

- Your target number is 12. Have your partner pick a number less than 12 and enter:

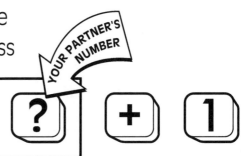

- Close your eyes and press $=$ $=$ ... until you think that you have hit the target number 12.

- Open your eyes to see if you hit the target.

Talk About It

What if you choose your own target number?

Could you play this target game using:

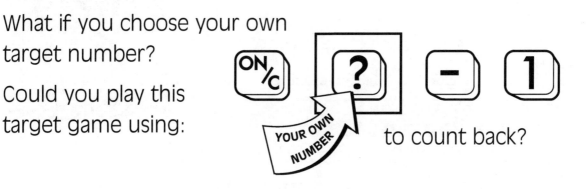

to count back?

100's Chart

If you mark the numbers your calculator shows,
what patterns do you see?

Work Together

- Use your 🖩 to count: ⓪ ➕ ② ⟌ ⟌ ⟌ ...
- Have your partner use a counter to cover each new number on the 100's Chart.
- Stop when you get to 50.
- Write a sentence that describes the pattern.
- Remove the counters and ⟨ON/C⟩ your 🖩.
- Try using ① instead of ⓪ and show the counting pattern on the 100's Chart.

Write About It

Compare the patterns you found.
- How are they alike?
- How are they different?

What If?

What if you counted by a number other than 2?

Try the activity again and use a number other than 2.

You'll need:
the 100's Chart and small counters.

100's CHART

1	2	3	4	5	6	7	8	9	10
11	12	13	14	15	16	17	18	19	20
21	22	23	24	25	26	27	28	29	30
31	32	33	34	35	36	37	38	39	40
41	42	43	44	45	46	47	48	49	50
51	52	53	54	55	56	57	58	59	60
61	62	63	64	65	66	67	68	69	70
71	72	73	74	75	76	77	78	79	80
81	82	83	84	85	86	87	88	89	90
91	92	93	94	95	96	97	98	99	100

Chart Patterns

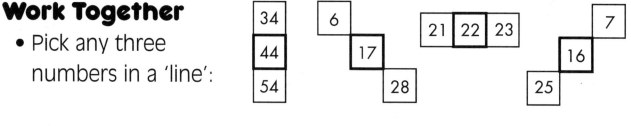

What sum patterns can you find for sums on the 100's and 200's Charts?

Work Together

- Pick any three numbers in a 'line':

34
44
54

6	
	17
	28

| 21 | 22 | 23 |

		7
	16	
25		

- How does the middle number compare to the sum of the other numbers? Use your 🖩 to help.

- Pick any square of numbers.

76	77
86	87

25	26	27	28
35	36	37	38
45	46	47	48
55	56	57	58

- Use your 🖩 to find patterns about sums.

125	126	127
135	136	137
145	146	147

Talk About It

Share the patterns you found.

Write About It

Write down 3 patterns you noticed.

You'll need:
the 100's and 200's Charts and small counters.

200's CHART

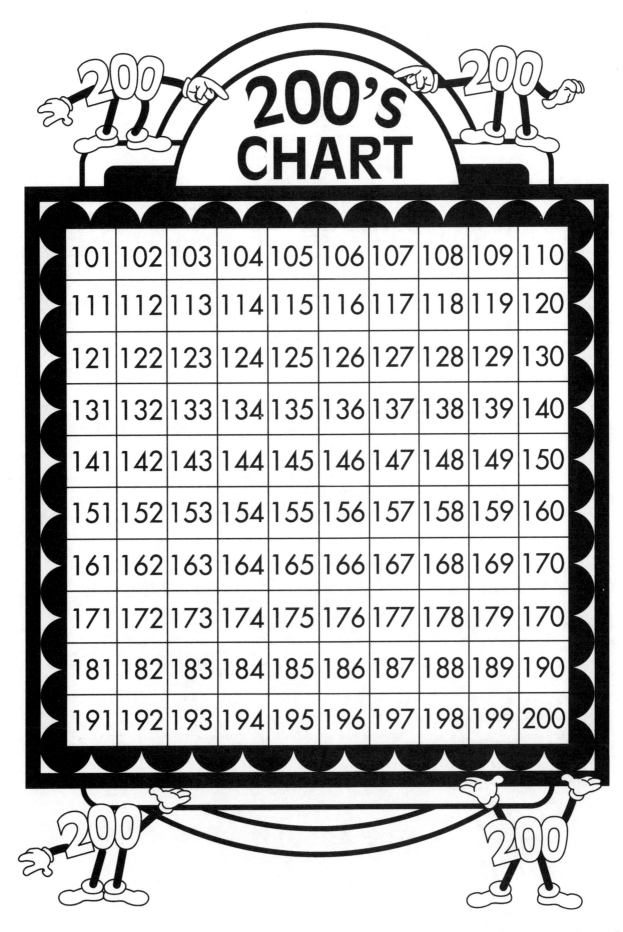

101	102	103	104	105	106	107	108	109	110
111	112	113	114	115	116	117	118	119	120
121	122	123	124	125	126	127	128	129	130
131	132	133	134	135	136	137	138	139	140
141	142	143	144	145	146	147	148	149	150
151	152	153	154	155	156	157	158	159	160
161	162	163	164	165	166	167	168	169	170
171	172	173	174	175	176	177	178	179	170
181	182	183	184	185	186	187	188	189	190
191	192	193	194	195	196	197	198	199	200

Guess My Rule: ➕ ➖

If you press 🟰 , can you guess my rule?

Work Together

• Without letting your partner see,
  pick a number for ❓ and enter:

ON/c ➕ ❓ 🟰 [any number] 🟰

(To enter your rule) (To hide your rule)

• Have your partner press: [any number] 🟰 to try to guess the rule.

– or –

• Press: [any number] 🟰 🟰 🟰 to find the rule.

• Try different rules.
  Sometimes use:

ON/c ➖ ❓ 🟰

Talk About It

How did you guess the rule?

Odd or Even?

When you ➕ or ➖ , can you predict when an answer will be odd or even?

Work Together

- Think about each puzzler before you use your calculator.
- Tell your partner what you think.
- Use your calculator to test numbers.
- Keep a record of your work.

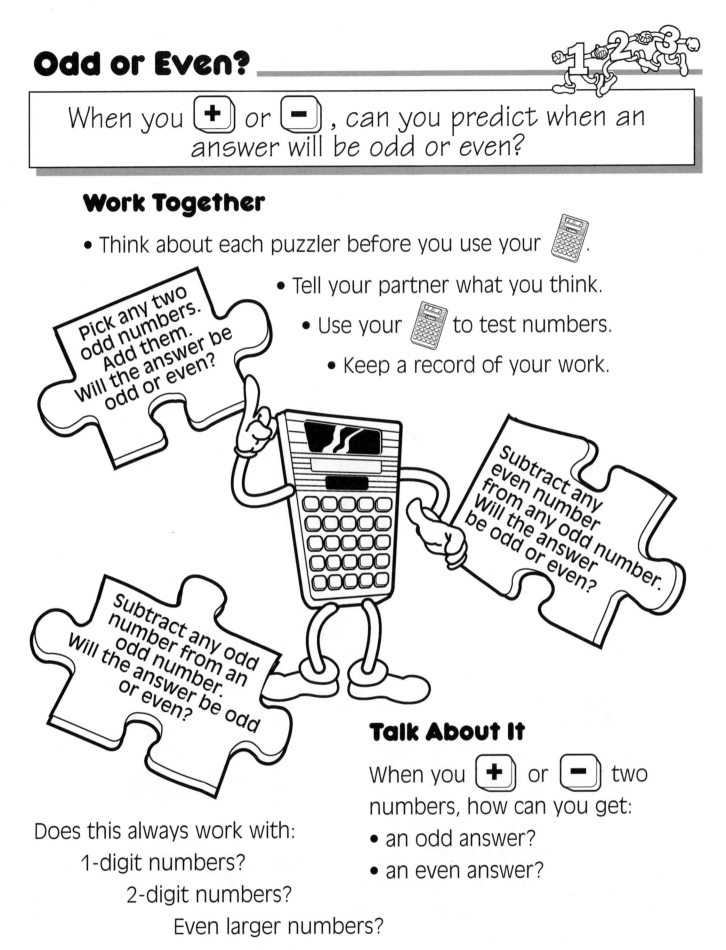

Pick any two odd numbers. Add them. Will the answer be odd or even?

Subtract any even number from any odd number. Will the answer be odd or even?

Subtract any odd number from an odd number. Will the answer be odd or even?

Does this always work with:
- 1-digit numbers?
- 2-digit numbers?
- Even larger numbers?

Talk About It

When you ➕ or ➖ two numbers, how can you get:
- an odd answer?
- an even answer?

My Turn, Your Turn: ⊞ ⊟

Can you predict if the answer will be odd or even?

Work Together

Take turns:

- Close your eyes to enter a number. Punch one or more number keys.

- Have your partner press:

- Predict if the answer will be even or odd.

- Press ⎡=⎤ to check.

  - Keep a record of your work.

    - Stop when you find 5 odds and 5 evens.

    - Repeat, but instead of using ⊞, use ⊟.

Write About It

What information did you use to predict if an answer would be even or odd?

Talk About It

Share your ideas with your classmates.

What did you learn about odd and even numbers?

Evens vs. Odds

When you ⊗ , can you predict when an answer will be odd or even?

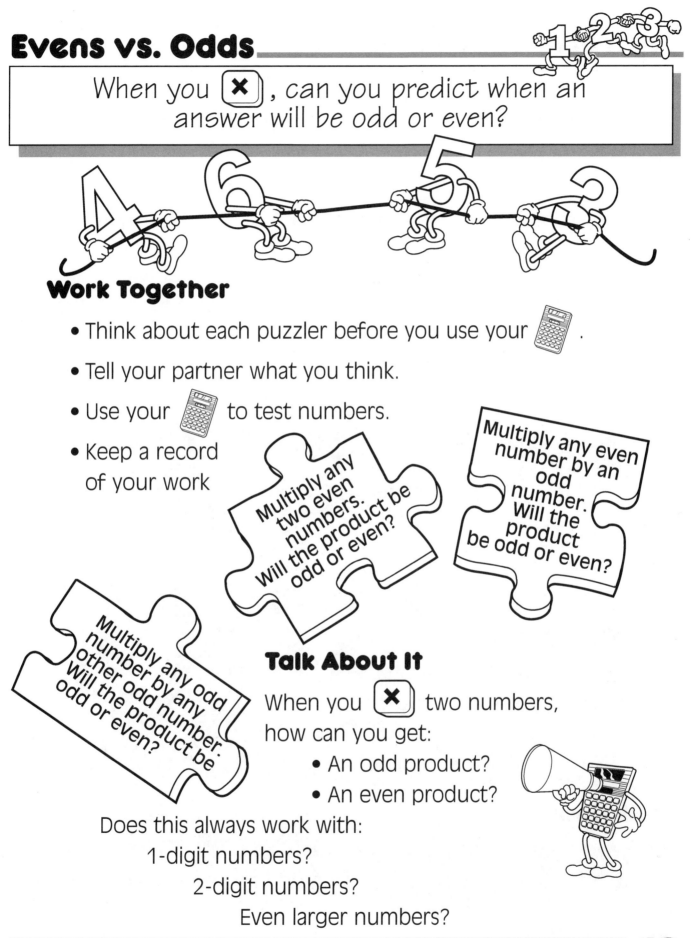

Work Together

- Think about each puzzler before you use your 🖩 .

- Tell your partner what you think.

- Use your 🖩 to test numbers.

- Keep a record of your work

> Multiply any two even numbers. Will the product be odd or even?

> Multiply any even number by an odd number. Will the product be odd or even?

> Multiply any odd number by any other odd number. Will the product be odd or even?

Talk About It

When you ⊗ two numbers, how can you get:

- An odd product?
- An even product?

Does this always work with:

1-digit numbers?

2-digit numbers?

Even larger numbers?

My Turn, Your Turn: ✗

> Can you predict if the product will be odd or even?

Work Together

Take turns:

- Close your eyes to enter a number. Punch one or more number keys.

- Have your partner press ✗ [any number].

- Predict if the answer will be even or odd.

- Press = to check.

- Keep a record of your work. Stop when you find 5 odds and 5 evens.

Write About It

What information did you use to predict if an answer would be even or odd?

Talk About It

Share your ideas with your classmates.
What did you learn about multiplying odd and even numbers?

Guess My Rule: [×] [÷]

If you press [=], can you guess my rule?

Work Together

- Without letting your partner see, pick a number for [?] and enter: [ON/C] [?] [×] [=]   [any number] [=]

  (To enter your rule)   (To hide your rule)

- Continue pressing: [any number] [=] as needed to find the rule.

- Have your partner press: [any number] [=] to try to guess the rule.

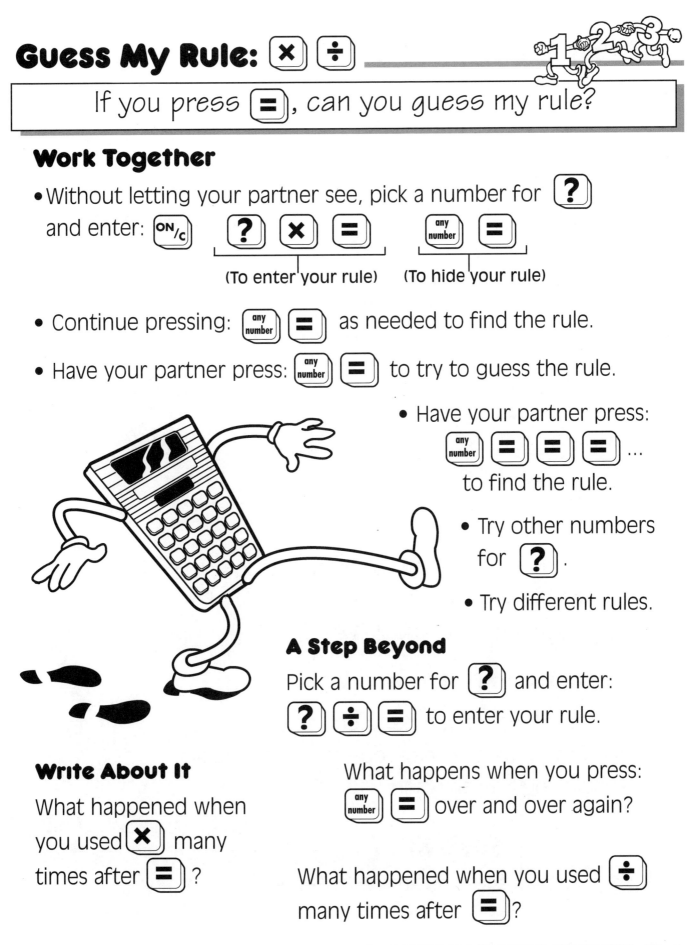

- Have your partner press:

  [any number] [=] [=] [=] ... to find the rule.

- Try other numbers for [?].

- Try different rules.

A Step Beyond

Pick a number for [?] and enter:

[?] [÷] [=] to enter your rule.

Write About It

What happened when you used [×] many times after [=]?

What happens when you press: [any number] [=] over and over again?

What happened when you used [÷] many times after [=]?

Patterns with Zeros

> Can you use patterns for "shortcuts?"

Work It Out

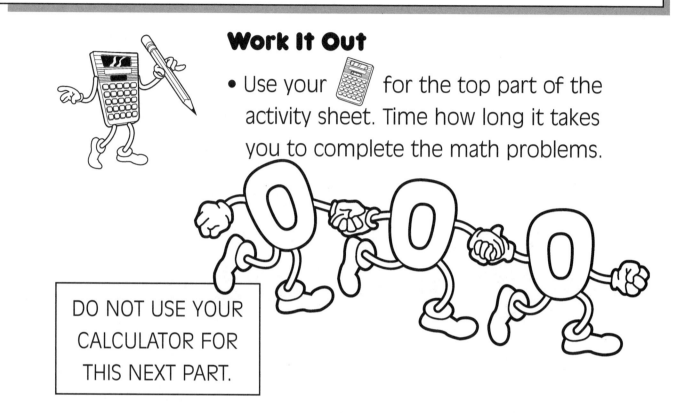

- Use your  for the top part of the activity sheet. Time how long it takes you to complete the math problems.

> DO NOT USE YOUR CALCULATOR FOR THIS NEXT PART.

- Complete the math problems on the bottom part of the Problem Sheet. Time how long it takes you to complete the math problems.

Talk About It

Did it take more time to do the top or the bottom part of the Problem Sheet? Explain. Did you find any "shortcuts?"

You'll need:

the "Patterns with Zeros" Problem Sheet.

PATTERNS WITH ZEROS PROBLEM SHEET

3	30	300
+4	+40	+400

6	60	600
+2	+20	+200

8	80	800
−2	−20	−200

$5 \times 3 =$ _____ $50 \times 30 =$ _____

$50 \times 3 =$ _____ $500 \times 3 =$ _____

$4 \times 2 =$ _____ $40 \times 20 =$ _____

$40 \times 2 =$ _____ $400 \times 2 =$ _____

$7 \times 2 =$ _____ $70 \times 20 =$ _____

$70 \times 2 =$ _____ $700 \times 2 =$ _____

2	20	200
+2	+20	+200

3	30	300
+6	+60	+600

7	70	700
−3	−30	−300

$4 \times 3 =$ _____ $40 \times 30 =$ _____

$40 \times 3 =$ _____ $400 \times 3 =$ _____

$6 \times 2 =$ _____ $60 \times 20 =$ _____

$60 \times 2 =$ _____ $600 \times 2 =$ _____

$5 \times 5 =$ _____ $50 \times 50 =$ _____

$50 \times 5 =$ _____ $500 \times 5 =$ _____

Section 2: **Number Sense**

OVERVIEW

The **Number Sense** activities encourage students to develop their intuitions about numbers and operations using estimation and mental computation. Through constructing and testing conjectures, students build on informal understandings and develop more systematic problem-solving strategies. The calculator's memory functions are introduced and featured in several activities.

TEACHING NOTES

Enough Money to Buy? Estimation is the focus of this activity. Have students cut out the bakery cards. Students predict whether they have enough money to purchase selected items at the bakery. Then they check their predictions with their calculator. Question students to elicit and encourage different strategies for estimating.

No More, No Less: Most students will benefit from working through this activity as a whole class before working with partners. This is one of several activities that involve a "target number." Beginning with a selected number entered into the calculator, students perform a sequence of operations to "reach the target." They are challenged to reach the target in 4 steps, that is, to use four operations beginning with the start number and ending with the target number. Encourage students to find more than one way to reach the target and remind them to record each "step" as a number sentence. After practicing in pairs, this activity could be presented as a class tournament in which teams challenge each other to reach a target in a given number of steps.

Hit the Target: This activity is appropriate for students beginning to develop operation sense with multi-digit numbers. Children should be encouraged to find and record many different ways of making each target number—using the calculator, if necessary, to check their thinking.

Two Tries: Four randomly selected digits are arranged to make two factors whose product is as close as possible to a target number. Encourage students to generate several number sentences to find the product that is closest to (or equal to) the target number. Observe and listen to groups as they work to assess students' use of number sense and estimation skills in determining which factors to use. The **Talk About It** discussion should focus on students' solution strategies. This activity can be modified to change the level of difficulty by using different target numbers, operations, or number of digits.

No! No! Keys: In this target activity, one operation and three digits are eliminated from use as students construct a number sentence to produce the target number. Encourage students to find more than one number sentence for a given target.

Yard Sale: This whole-class activity introduces students to the calculator's memory functions. Students keep track of how much money they have left after buying items at a yard sale. Students will find it necessary to use decimal notation when entering amounts of money into the calculator, and to interpret a numerical display of say, 0.4 as 40¢. In the *Talk About It* section, students consider the role of the memory recall key and discuss what the numbers in the display indicate.

Go For It! This activity features the calculator's memory functions. Students devise a winning strategy for reaching a target number with their calculator. Keeping a written record of each player's moves will assist students in developing strategies. Many students find it helpful to "work backwards" as they plan a series of turns.

Start to Finish: In this problem-solving activity, pairs of students use estimation to determine which path along a maze will produce the most number of points. During class discussion students share estimation strategies and justify the selection of certain solution paths. Have students create their own mazes.

Enough Money to Buy?

> Is there enough money to buy what is listed on the Bakery Cards?

Work Together

- Read a Bakery Card.

- Predict if there is enough money to buy what is listed.

- Use your 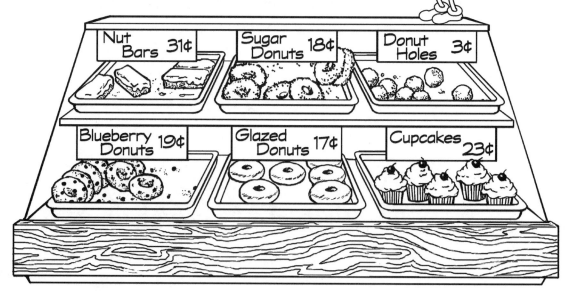 to check as needed.

- Make up your own Bakery Cards. Trade with classmates.

Write About It

What estimation strategies did you use? Describe one or two you liked best.

| Nut Bars | 31¢ | Sugar Donuts | 18¢ | Donut Holes | 3¢ |
| Blueberry Donuts | 19¢ | Glazed Donuts | 17¢ | Cupcakes | 23¢ |

You'll need:

the "Bakery Cards" page.

No! No! Keys

What keys can you punch so the  displays the target?

Work Together

- Pick four "No! No!" keys from below that you will not use:

  Cross off one of these keys:

  $+$ $-$ $\times$ $\div$

  Cross off three of these keys:

- Pick a target number from the following numbers: **50, 63, 100, 175**

- Without using a "No! No!" key, try to make your  show the target number.

Talk About It

What keys worked for different target numbers? How did you decide?

Write About It

Record the keystrokes you used. Try different ways to reach your target number. Choose a new target number and try again.

Yard Sale

How much money will you have left?

Work Together

- Press [ON/C] to clear the memory.

  - From the amounts listed, choose how much money you will take to a yard sale:
    $2.00, $3.00, $5.00

- Enter "your amount" and then [M+] into your 🖩.

- Pick something to "buy".

  - Enter its price and press [M-].

  - Pick a few other items and enter their prices as you did above.

    - Press the [MRC] key to see how much money you have left.

Talk About It

When you press [MRC], what does the number in the display window represent? How is [M+] different from [M-]?

©Learning Resources, Inc.

Go For It!

Is there a winning strategy to hit the target number?

Work Together

- Pick a target number from the following numbers: **15, 21, 35**

- By adding in the memory, try to reach this target without going over.

- Press [ON/C] to clear the memory.

- Take turns entering numbers 1 through 5, followed by the [M+] key.

- Press [MRC] to find the sum of all the numbers put into memory.

Write About It

Is there a strategy for reaching the target number first?

What is the least number of turns to reach the number?

What is the most number of turns?

Talk About It

Share what you found.

Start to Finish

What path to the Finish gets you the most points?

Work Together

- Study the number maze.

- Predict a path that will give you the most points.

- Always go downward, left, or right, never back up.

- Add or subtract in your head as you go.

- Use your to help or check.

- Record the points of the path you choose.

- Do the maze again. Try for the least points.

Write About It

How did you choose your path?

What strategies did you use?

Talk About It

Compare your path with your partner.

You'll need:
the "Start to Finish Maze."

START TO FINISH MAZE

START HERE

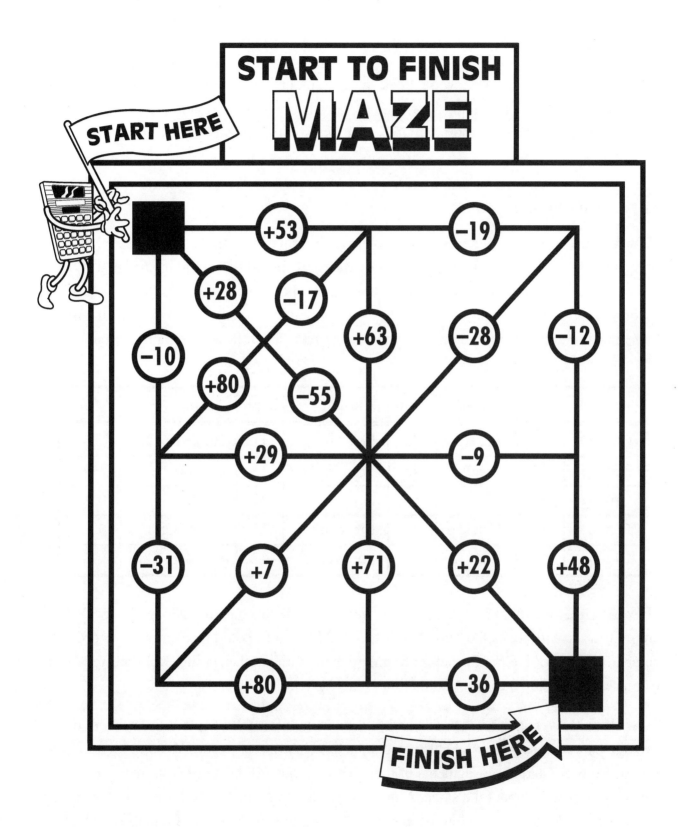

+53 −19

+28 −17

−10 +63 −28 −12

+80 −55

+29 −9

−31 +7 +71 +22 +48

+80 −36

FINISH HERE

Section 3: **Measurement**

OVERVIEW

The activities in this section involve linear measurement, using both English and metric systems. Measurements of perimeter and area are also addressed. Estimation and mental computation are emphasized in most activities. The concept of "average" is explored informally.

TEACHING NOTES

How Much Closer? Students estimate and measure, in centimeters, a series of line segments representing two paths. Mental computation should be encouraged at first. Then, use a calculator to total the lengths of the line segments or to compute the "actual" distance according to the given scale. In *A Step Beyond*, students are challenged to determine how long it would take different animals to travel the paths. The following data is presented as a reference:

Animal	Mph	Animal	Mph	Animal	Mph	Animal	Mph
Cheetah	70	Elk	45	Pig	11	Human	25
Zebra	40	Coyote	43	Wild Turkey	15	Greyhound	39
Grizzly	30	Lion	40	Grey Fox	42	Garden Snail	0.03
Rabbit	35	Chicken	9	Elephant	25	Giant Tortoise	0.17
Giraffe	32	Cat	30	Squirrel	12	Reindeer	32

How Much Closer Now? In this activity, pairs of students construct new paths for the animals in the activity called *How Much Closer?* Students should measure and record the actual distances of their paths before trading papers with another group. Remind students to estimate before they measure. Estimation is the focus of this exercise.

How Does It Average Out? In this activity, students informally explore the concept of average by comparing their heights using square inch strips of paper. Working in groups of six, students cut and paste segments of their strips so that all the heights in their group "average out" to be about the same. A measuring tape or other device for measuring height in inches should be displayed in the classroom. Also, post a class list of names on which students can record their heights to the nearest inch.

Clip Drop: A game in which students measure the distance, in centimeters, between two paper clips provides another context for exploring the concept of average. This activity follows the same format as the activity called *How Does It Average Out?*

How Many Steps? After determining the "average" number of steps required to walk 10 yards, students use this information to estimate distances between places in the school or on the playground. Although averages should be computed for groups of 6 to 8, students can work in pairs or individually to find the number of steps in 10 yards.

This activity could be conducted over two class periods—collecting data one day and determining averages the next. It is likely that the average number of steps for many groups will involve decimals. Promote discussion about the meaning of decimals in terms of number of steps can be modified to change the level of difficulty by using different target numbers, operations, or number of digits.

A Longer Perimeter: Students construct tangram puzzles having different perimeters. Listen to students as they work to assess their use of strategies for making a perimeter longer or shorter. For example, some students might conclude that a longer perimeter can be made by exposing more sides of the tangram pieces rather than having sides connect to form the interior of a puzzle.

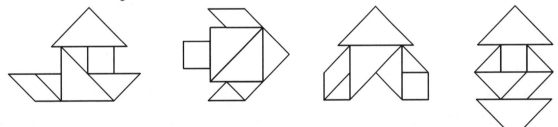

How Many? Use this activity prior to addressing the concept of area. As students determine how many dots are in each region, strategies such as finding the number of dots in a row and multiplying by the number of rows may emerge.

A. 13 x 24 = 312   C. 14 x 35 = 490   E. (13 x 10) + (23 x 12) = 406
B. 19 x 19 = 361   D. 43 x 30 = 1290

What Area is Covered? The concept of area is explored informally as students determine how many square units cover a given region.

A. 10 x 10 = 100   D. (30 x 20) + (20 x 15) = 900
B. 12 x 30 = 360   E. (20 x 10) + (15 x 20) + (20 x 10) = 700
C. (30 x 30) + (20 x 10) = 1100   F. (20 x 30) – (10 x 15) = 450

How Much Closer?

Which animal is closer to water? How many miles closer?

Work Together

- Measure each piece of the elephant's path to water in centimeters.
- Mentally add the lengths of each part of the path.
- Have your partner measure and add the lion's path.
- Check your addition estimates with your .

Write About It

Using the scale 1 cm = 3 miles, tell how far each animal must travel to find water.

How do the distances compare?

A Step Beyond

Research typical travel speeds of animals.

About how long might it take different animals to travel the paths to the water?

You'll need:
the "Animal Paths" map and a centimeter ruler.

ANIMAL PATHS

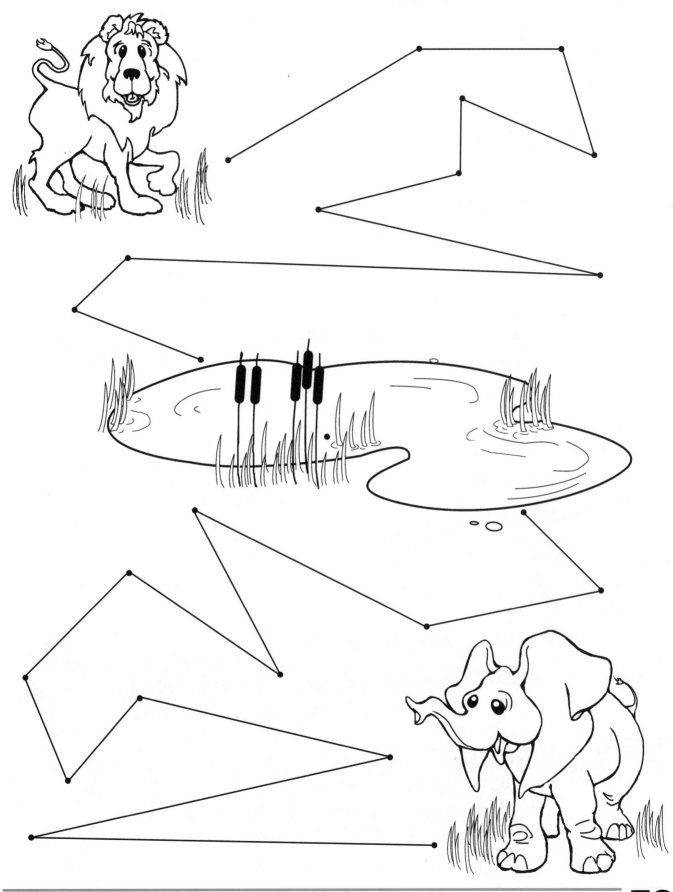

How Much Closer Now?

Which animal *do you estimate is* closer to the water?

Work Together

- The elephant and the lion must find new paths to the water.

- Draw a new path for each animal.

- Measure each path in centimeters and tell the scale:

    1 cm = **?** miles.

- Trade papers with another group.

- Estimate which animal's path to the water is shorter.

- Measure both paths. Mentally add the lengths of each part of the paths or use your  to help.

- Using the scale from above, how far must each animal travel to water?

Talk About It

Explain what strategies you used to estimate the lengths of the paths.

You'll need:

the "Create Your Own Animal Paths" map, and a centimeter ruler.

©Learning Resources, Inc.

CREATE YOUR OWN ANIMAL PATHS

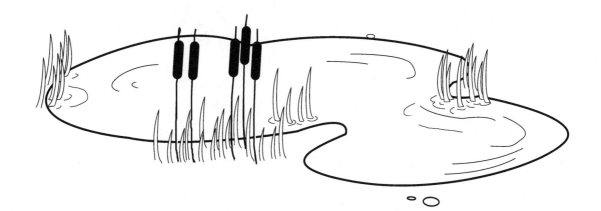

How Does It Average Out?

What's a "typical" height for someone in your class?

Work Together

- Measure your height in inches.
- Record on the class chart.
- Cut and tape square inch strips together to show your height.
- Join with other students to form a group of six.
- Compare your strip to others in your group.
- Cut inch squares off longer strips and tape them to the shorter ones, until all of the strips average out to be "about the same."

Talk About It

Use a ▨. Find the sum of the heights of all in your group. Then divide by the number in the group. That is the **average**.

How does your work with the ▨ compare to what you did with the strips?

 You'll need:
tape, square inch paper, and a measuring tape.

SQUARE INCH PAPER

Clip Drop

What is the "average" distance between paper clips in the Clip Drop Game?

Work Together

- Play the "Clip Drop" game.

- Cut centimeter strips to show how far apart the clips landed.

- Compare the lengths of the 10 strips.

- Cut centimeters off longer strips and tape them to shorter ones until all strips average out to be "about the same."

- Use a . Find the sum of the distances on the "How Far Apart?" table. Divide by 10.

Talk About It

How does your work with the  compare to what you did with the strips?

 You'll need:
two paper clips, tape, "Clip Drop Game," and the "Square Centimeter Paper."

CLIP DROP GAME

Take turns

- Hold two paper clips in one hand, waist high.

- Drop them.

- Have your partner measure the distance between the clips to the nearest centimeter.

- Record the measurements on the table below.

- Repeat the drop 9 more times to fill the chart.

How Far Apart?

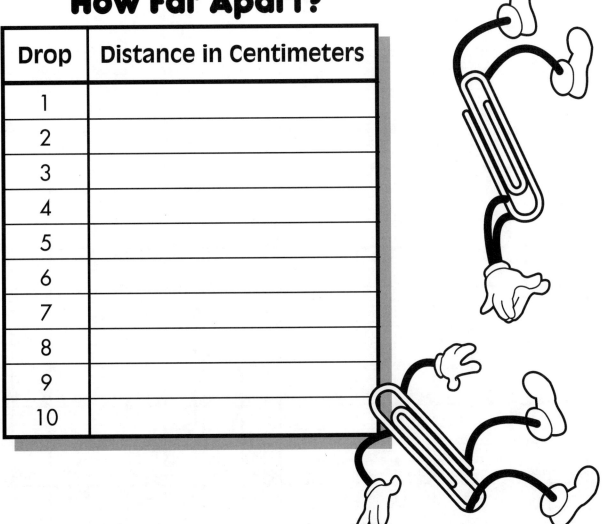

Drop	Distance in Centimeters
1	
2	
3	
4	
5	
6	
7	
8	
9	
10	

SQUARE CENTIMETER PAPER

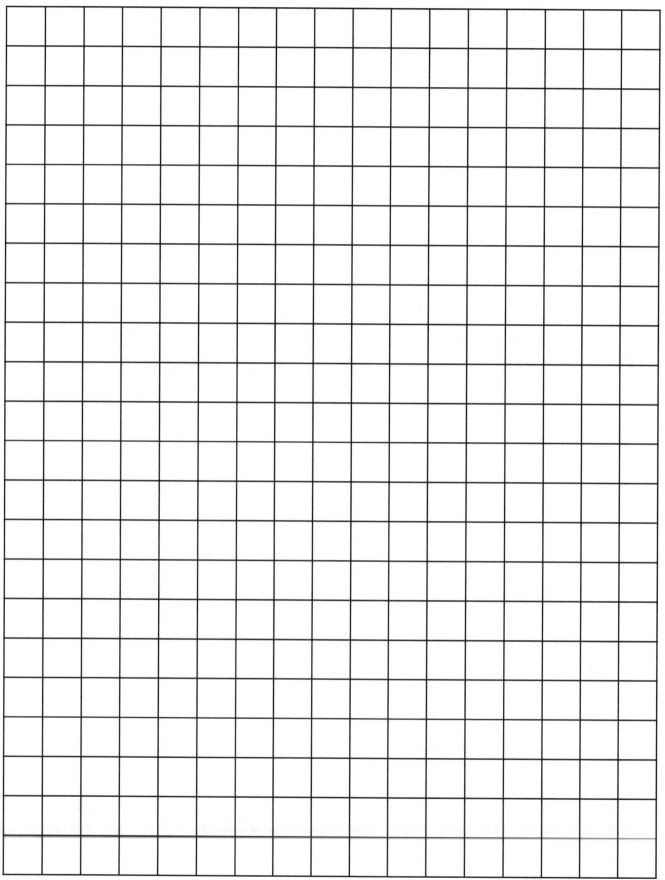

How Many Steps?

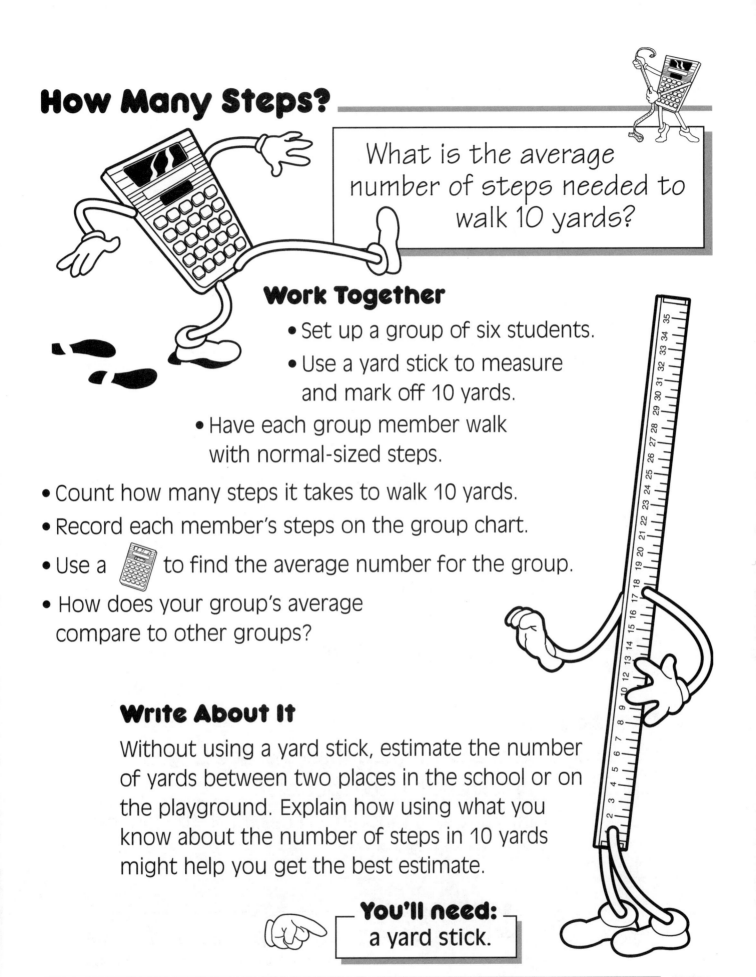

What is the average number of steps needed to walk 10 yards?

Work Together

- Set up a group of six students.
- Use a yard stick to measure and mark off 10 yards.
- Have each group member walk with normal-sized steps.
- Count how many steps it takes to walk 10 yards.
- Record each member's steps on the group chart.
- Use a ▣ to find the average number for the group.
- How does your group's average compare to other groups?

Write About It

Without using a yard stick, estimate the number of yards between two places in the school or on the playground. Explain how using what you know about the number of steps in 10 yards might help you get the best estimate.

You'll need: a yard stick.

A Longer Perimeter

Which tangram puzzle shape has the longer perimeter?

Work Together
- Cut out the 7 tangram pieces.
- Solve one of the tangram shapes.
- Glue the shape puzzle on paper and find its perimeter in centimeters.
- Try to make a second shape puzzle with a longer perimeter.

Talk About It
Did you have a plan for making a longer perimeter?

A Step Beyond
- Using only the triangles, which shape has the longest perimeter? Which shape has the shortest?
- Using different tangram pieces, find the longest and shortest perimeters possible.

You'll need: scissors, glue, tangram pieces, and a ruler.

TANGRAM PUZZLE PIECES

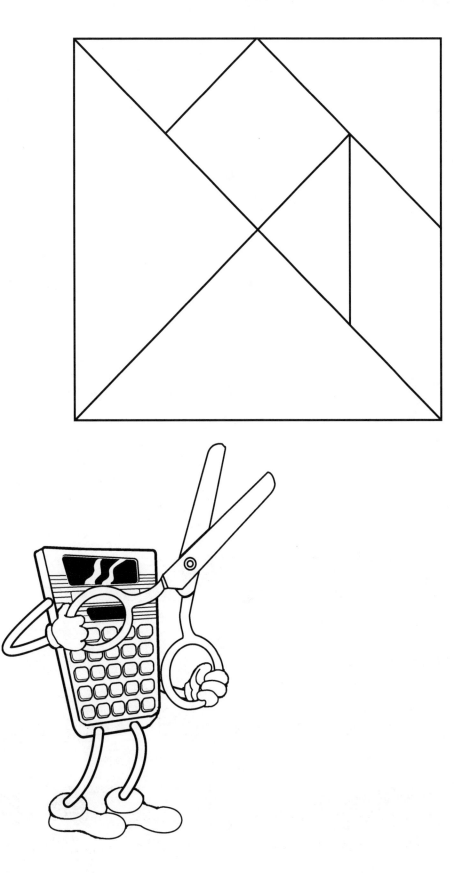

How Many?

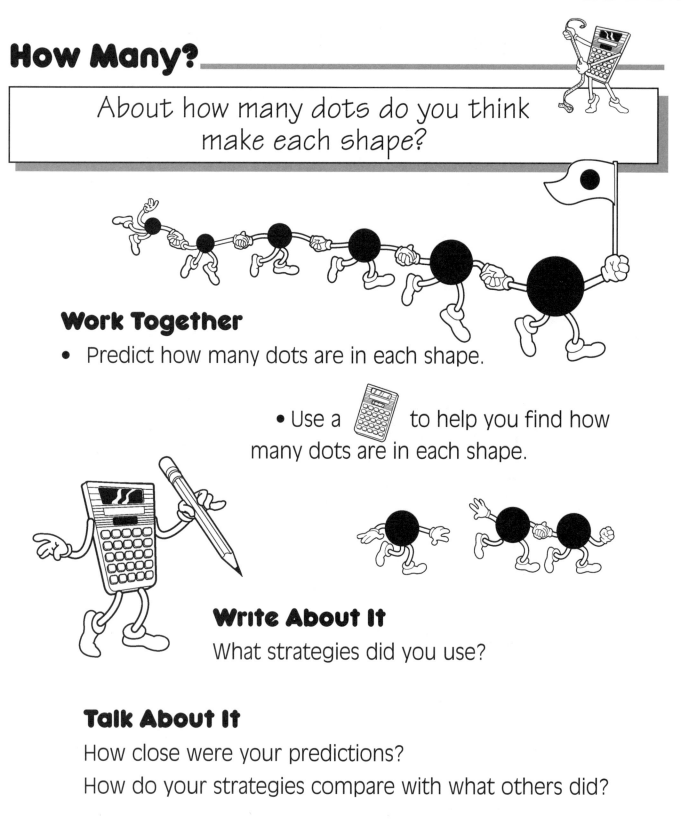

About how many dots do you think make each shape?

Work Together

- Predict how many dots are in each shape.

- Use a 🖩 to help you find how many dots are in each shape.

Write About It

What strategies did you use?

Talk About It

How close were your predictions?

How do your strategies compare with what others did?

You'll need:
the "Dot Shapes" work sheet

DOT SHAPES

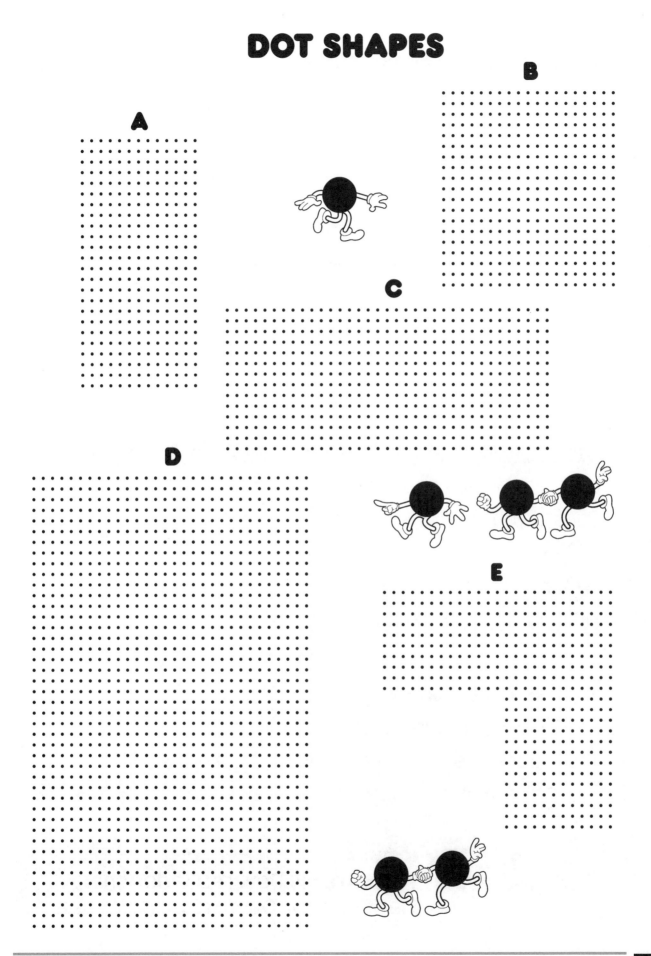

What Area is Covered?

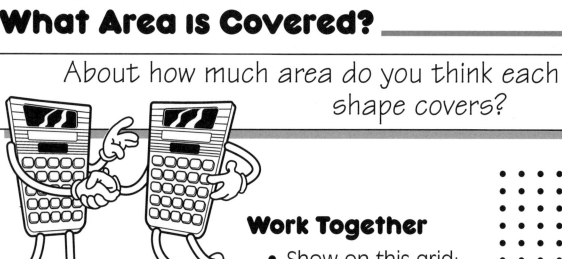

About how much area do you think each shape covers?

Work Together

- Show on this grid:

  1 unit of length

  1 unit of width

  1 square unit of area.

- Use your to find how many square units are covered by each shape on the grid sheet.

- Record your findings.

Talk About It

How did you find the area of each shape?

A Step Beyond

Use your own shapes to cover the dots.

Trade with classmates and find the areas.

You'll need:

the "What Area is Covered?" dot chart and the "Use Your Own Shapes" dot chart.

What Area is Covered?

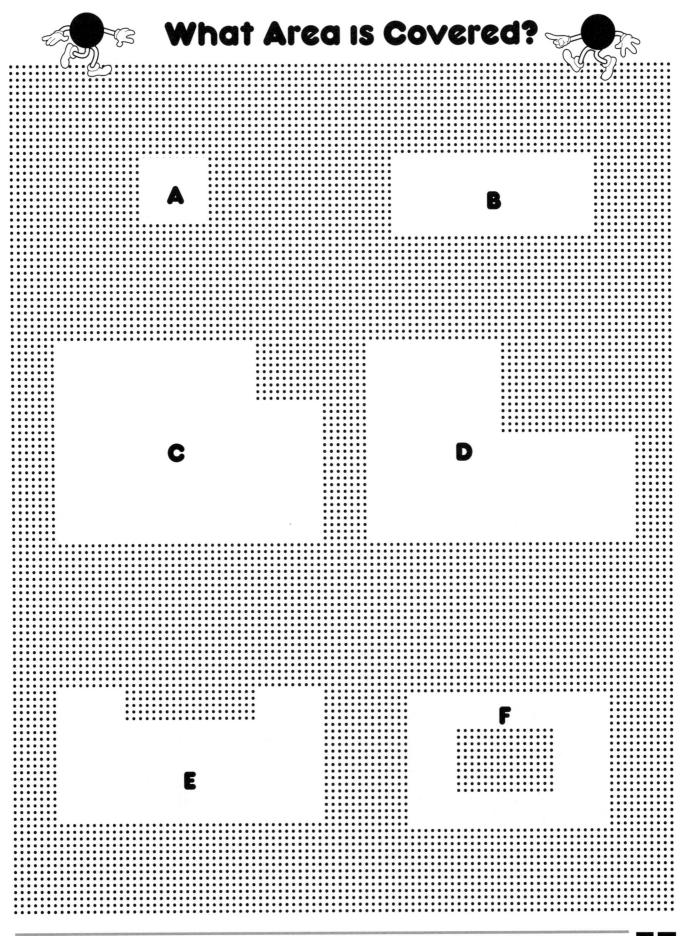

Use Your Own Shapes

Section 4: **Data and Chance**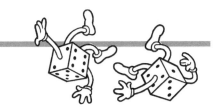

OVERVIEW

The activities in this section focus on collecting, displaying and interpreting data. Students consider notions of chance in terms of predicting outcomes based on their interpretations of data collected from repeated trials of an experiment.

TEACHING NOTES

Mystery Shake: In this activity, students consider whether it is better to make predictions based on large or small samples of data. Each pair of students will need a paper lunch bag with one corner snipped (just enough to allow a cube to be visible without falling through). Bags should contain 10 cubes, 8 of one dark color and 2 of one light color. Staple or seal bags in some other way so that students cannot look inside until they are instructed to do so.

As a whole-class activity, students discuss ways of finding the average number of light and dark cubes. After all ideas have been shared, have someone actually rearrange the cubes to "even them out" as a way of showing the average.

About How Many? Students work in pairs to collect data about light and dark cubes. These data are recorded numerically on a class chart and collections of cubes are displayed on the chalk ledge. If it is difficult to stack the cubes, arrange them in separate piles on a table.

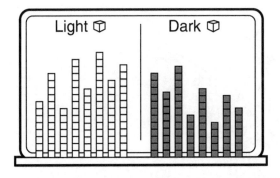

Team	Light 🎲	Dark 🎲
1		
2		
3		
4		
5		
6		

Who is Closer? Students look for trends in performance data to indicate whether or not there is improvement with practice. Each pair of students will need a dictionary or other "large" book.

How Many Spaces? In this activity students encounter a situation in which finding an average is not necessarily the most appropriate solution to the problem. Students use tallies to record the number of letters in their last names. A frequency chart, like the one below, should be displayed on the chalkboard or overhead projector for students to record their data.

Number of Letter	Tally	Frequency
3		
4		
5		
6		
7		
8		

Computing the class average for number of letters in the last name will require multiple steps and may warrant whole-class discussion before students are instructed to *write about it*.

How Close? In this whole-class activity, students close their eyes and estimate when one minute has elapsed. As you time one minute, record the number of seconds (in 5 second intervals) on the overhead. When students open their eyes, the last number displayed will indicate their "estimate." Have students use tallies on a frequency chart to record the number of seconds their eyes were closed. Use this data to compute the class average.

Mystery Shake

10 cubes of 2 colors are in a bag.
After 10 shakes, can you predict how
many of each color are left in the bag?

Work Together

- Take turns shaking the bag.
- Peek at the cube sticking out.
- Tally the color.
- After 10 shakes, predict the number
  of light and dark cubes in the bag.

- Repeat the activity with a new tally.
  Do you want to revise your prediction?

- On the class chart,
  record how many
  light and dark cubes
  are sticking out of the bag.

- Use your to
  find the average
  for each color.

- Open your bag and count the light and dark cubes.

Write About It

Which is best, a prediction based on
your team's data or a prediction
based on the class data?
Explain your thinking.

You'll need:
10 cubes of 2 colors, and a paper lunch bag.

About How Many?

Do you think there will be more light or more dark cubes after 20 spins of the Color Chooser?

Work Together

- Take turns spinning the Color Chooser.
- Take a cube indicated by the Chooser.
- Stop when you have 20 cubes.

- On a class chart, record how many light and dark cubes you have.

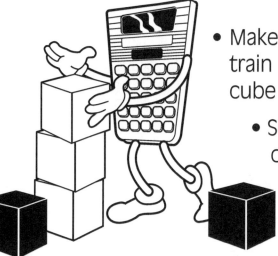

- Make a light cube train and a dark cube train.
  - Set your cube train on a chalk ledge.

Talk About It

What could you do to find the average number of light cubes:

-by moving cubes?

-by using a calculator?

How would you find the average number of dark cubes?

You'll need: the "Color Chooser," and about 10 light and 15 dark cubes.

©Learning Resources, Inc.

COLOR CHOOSER

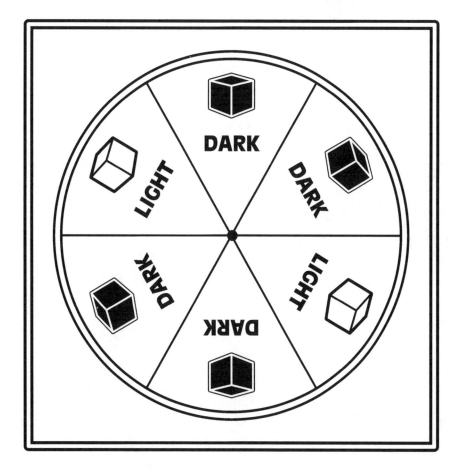

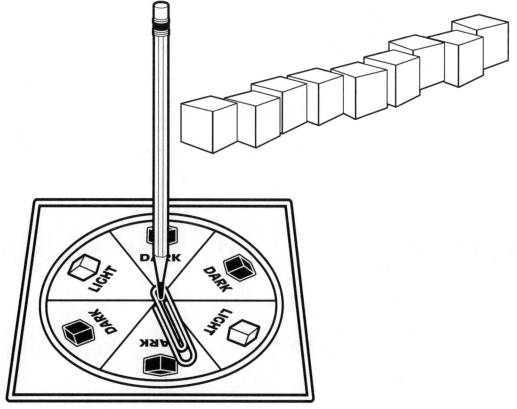

Who Is Closer?

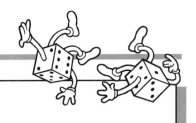

Do you improve with practice?

Work Together

- Find a large book with many pages, such as a dictionary.
- Find out how many pages are in the book to find your range of numbers.
- Pick a "target" page number for your partner and record it.
- Have your partner choose a "target page" for you.

- Without flipping the pages, try to open the book to the target page.
- Record the page number on the chart.
- Use your  to find how close you came to the page.

Talk About It

Did you improve with practice? How do you know?

What If?

Would you expect to see the same pattern in your data if you were to play the game again?

You'll need:

a "large" book and the "Data Charts" worksheet.

Data Charts

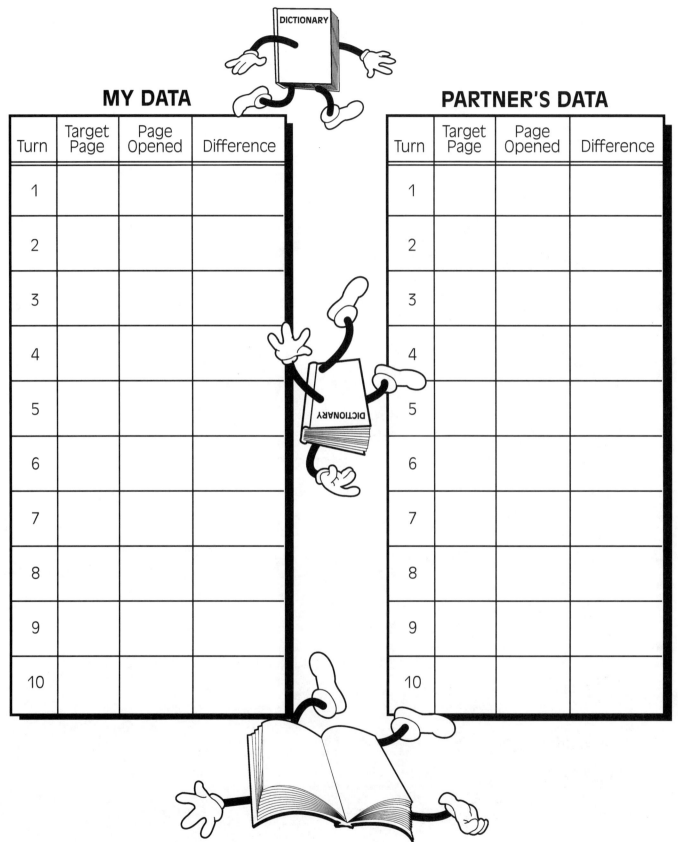

MY DATA

Turn	Target Page	Page Opened	Difference
1			
2			
3			
4			
5			
6			
7			
8			
9			
10			

PARTNER'S DATA

Turn	Target Page	Page Opened	Difference
1			
2			
3			
4			
5			
6			
7			
8			
9			
10			

How Many Spaces?

About how many spaces for a last name should be on a school form?

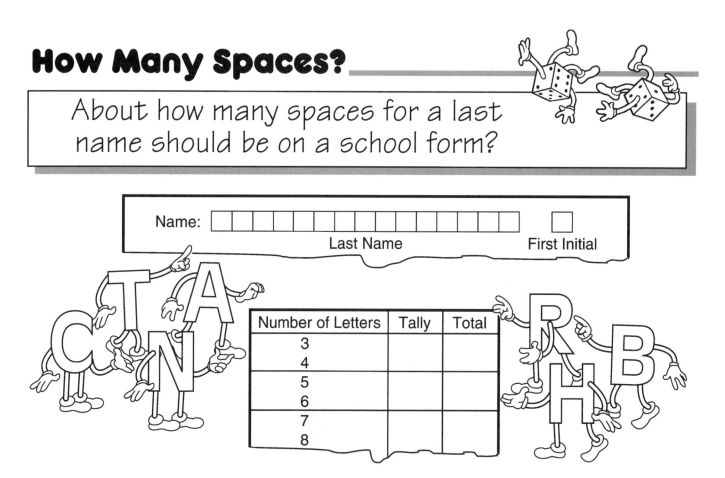

Name: ☐☐☐☐☐☐☐☐☐☐☐☐☐☐☐ ☐

Last Name   First Initial

Number of Letters	Tally	Total
3		
4		
5		
6		
7		
8		

Work Together

- Count the letters in your last name.
- Record the number on the class chart with a tally mark.
- When the tally column is complete, fill in the total column.

Write About It

What is the average number of last name letters in your class?

Talk About It

How would you answer the name space problem?

Did you use an average? Why or why not?

©Learning Resources, Inc.

How Close?

Can you estimate when one minute is up?

Work Together

- As a class, close your eyes when your teacher says, "Now!" Open them when you think one minute is up.

- Write down the last number the teacher has written on the overhead that tells, to the nearest five seconds, how long your eyes were closed.

- As your teacher directs, fill in the tally column on the class chart.

- When the tally column is complete, fill in the total column.

Write About It

On the average how long did the class keep their eyes closed? Use your .

Talk About It

How close to one minute was the class average?

Would you expect estimates to improve if the experiment is repeated? Try it!

Four-Function Calculator

AWARD CERTIFICATE

To:

for excellent work
in solving calculator problems

Teacher

Date